This Walker book belongs to:

Kinds of Clocks

There are two different types of clocks:

ANALOGUE

This kind of clock has hands which move as time goes by. Around the edge are numbers, with "12" at the top and "1" just to the right.

DIGITAL

Electronic devices like phones and computers usually have a digital clock. This kind of clock shows the exact time in numbers.

It's easier to see the time going by on an "analogue" clock ...

and that's exactly what we're going to do in this book!

For Ya-Ling Huang, with love and many thanks xxx V. F.

For my adorable niece Rhea and her family Y-L. H.

First published 2024 by Walker Books Ltd, 87 Vauxhall Walk, London SE11 5HJ • Paperback edition published 2025 • 10 9 8 7 6 5 4 3 2 1

This book has been typeset in Anke Sans

Printed in China

British Library Cataloguing in Publication Data: a catalogue record for this book is available from the British Library

Hardback ISBN 978-1-5295-0100-1 • Paperback ISBN 978-1-5295-2322-5 • www.walker.co.uk

Telling the Time with Anna

Vivian French

illustrated by

Ya-Ling Huang

WALKER BOOKS
AND SUBSIDIARIES
LONDON • BOSTON • SYDNEY • AUCKLAND

"When's Zane coming for tea?" Anna asked. "Is it soon?"

Grandpa laughed. "You asked me that 2 minutes ago! Why don't I show you how to tell the time yourself?"

12 1 2 3 4 5 6 7 8 9 10 11
Calendar
July
Mon Tue Wed Thu Fri Sat Sun
1 2 3 4 5 6 7
8 9 10 11 12 13 14
15 16 17 18 19 20 21
22 23 24 25 26 27 28
29 30 31

"You remember how I measured you and Whiskers?" he said.

Anna nodded. "You marked how tall we were on the wall.

Whiskers' tail was 20 centimetres!"

"Well," said Grandpa, "we can measure time, too..."

"Every day has a beginning, a middle and an end," he continued. "You start your day when you wake up. In the middle, you have lunch, and at the end, you go to bed.

But you can split the day into even smaller parts. Every day has **24 hours** ...

and each hour has **60 minutes** ...

and each minute has **60 seconds**."

Grandpa took the clock from the wall. "Let's watch time go by!"

Anna peered at the clock: it had three pointers.

"The red stick is moving, but the blue ones aren't."

"The sticks are called 'hands'," Grandpa told her, "and actually, they're *all* moving. But the red hand measures SECONDS, and seconds whizz by very quickly."

The hands on a clock move in a circle: right from the top, then back up in a loop... We call this CLOCKWISE!

Anna watched the "second" hand...

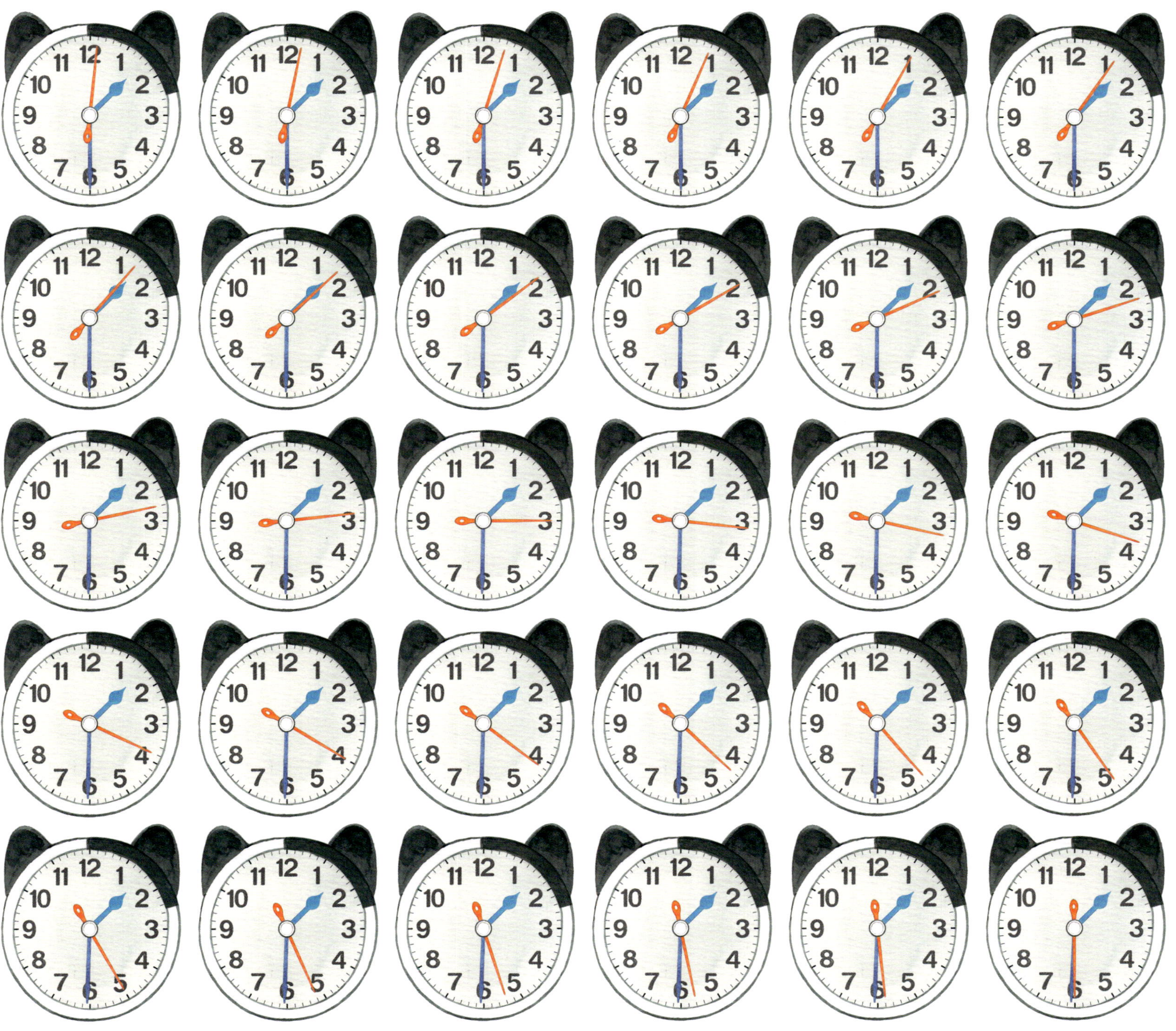

"It's gone halfway round the circle already!" she said.

"When the red hand is near 12, see what this blue MINUTE hand does," said Grandpa.

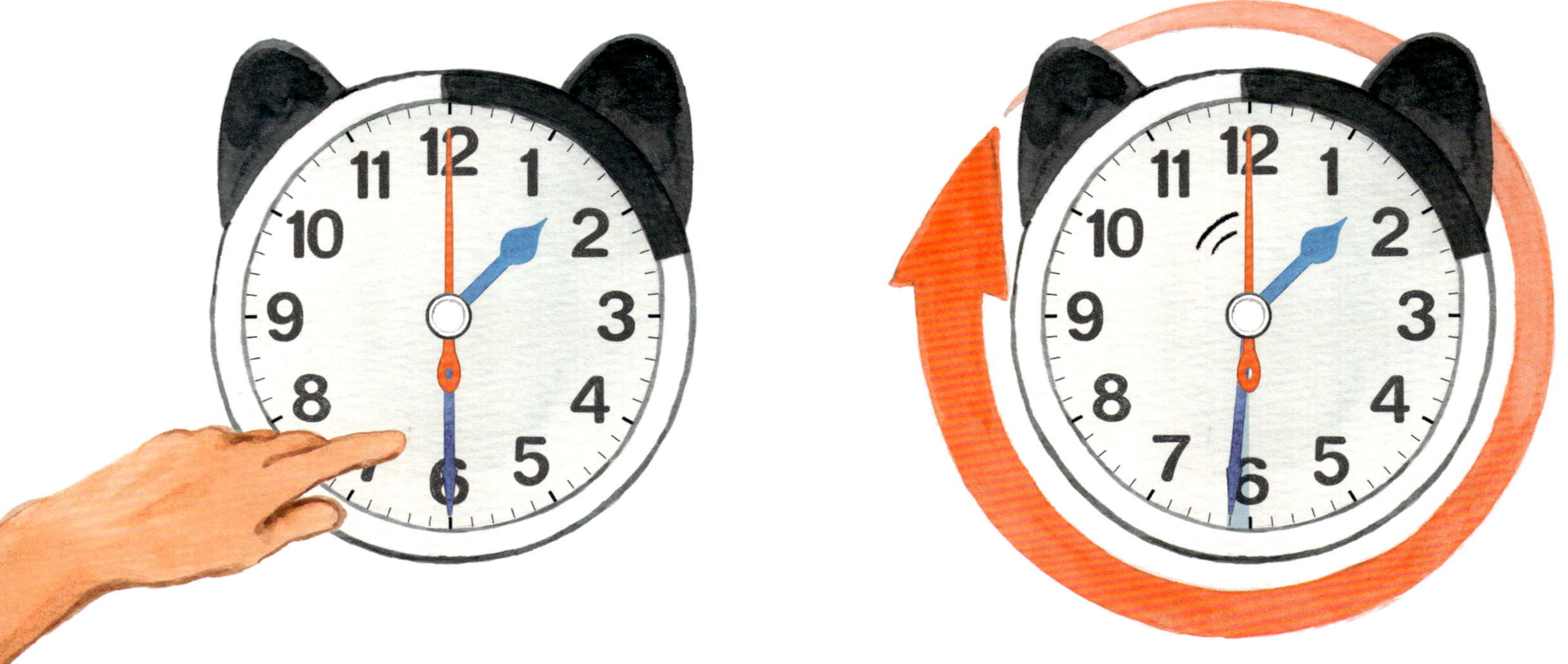

"OOOH! It did a little jump!" Anna was excited.

"That's one whole minute," said Grandpa. "Let's see if we can count another. 1 to 60: here we go! 1, 2, 3..."

When they reached 60 seconds – hop! – the minute hand jumped again.

"And there are 60 minutes in an hour," said Grandpa. Anna looked hopeful. "Can we count 60 minutes, too?" Grandpa laughed. "We might get terribly bored. Now, count how many seconds it takes me to fetch some paper!"

Remember! 60 seconds is 1 minute and 60 minutes is 1 hour.

Anna had only counted to 20 when Grandpa came bustling back.

"Let's make our own clock!" he said. "I'll draw the circle, and you put in the numbers."

"We need a short hand to point at the hours and a long hand to point at the minutes... There! Now we've got our practice clock all ready to go!"

He moved the hands to 12. "What time is it?"

"12 and no minutes!" Anna was very pleased with herself.

"Perfect!" said Grandpa. "You're ready for the next part."

Anna could also have said "12 o'clock", or just "12". "O'clock" means "of the clock"!

He drew a larger circle, and put in another row of numbers.

"Now I'm adding groups of 5 minutes," he said.

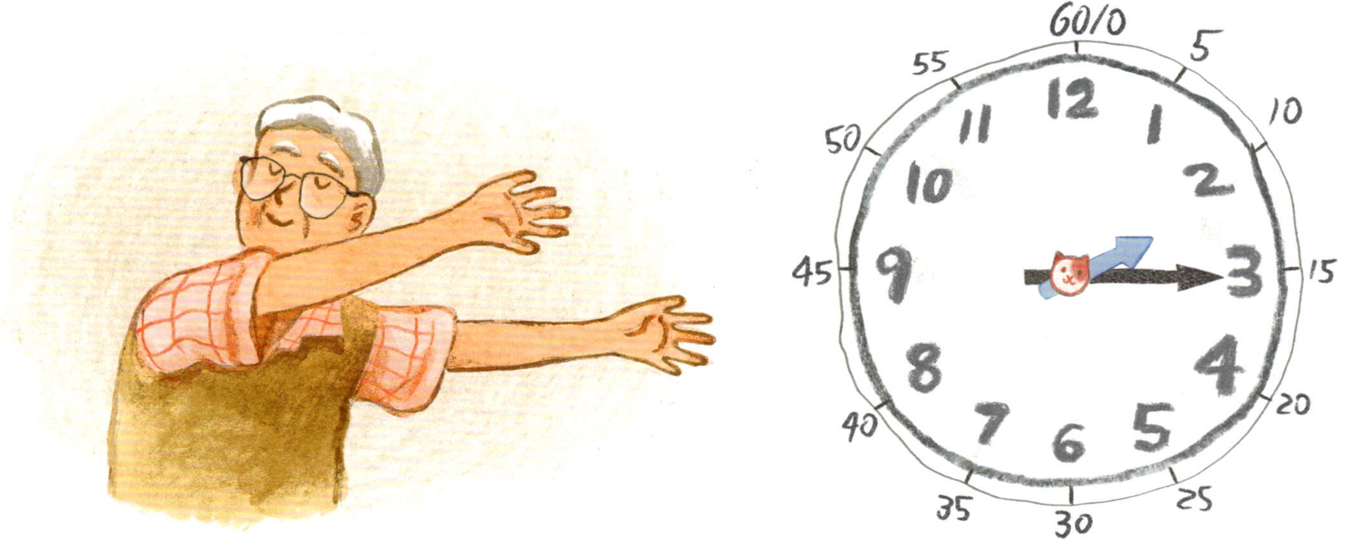

"When the clock looks like this, it's 15 minutes past 2."

Grandpa moved the hands again. "What time is it now?"

"30 minutes past 2!" said Anna.

"Exactly!" Grandpa clapped.

"Here's a tricky bit..."

30 minutes past the hour is also called "half past two", because the minute hand is halfway round the circle. 15 minutes past is a quarter of the way, so it's also "quarter past"; 45 minutes is three quarters of the way, with just one quarter left, so it's also "quarter to".

"The clock only goes to 12," said Grandpa, "but there are 24 hours in the day. So, we have to count to 12 TWICE: once in the morning, then again in the afternoon and evening. We call morning times 'a.m.' and later ones 'p.m.'."

"Hmmm ... 'a.m.' and 'p.m.'," Anna said to herself.

"Right then," said Grandpa. "Let's practise. What's the time, Anna?"

When they'd finished practising, Grandpa got the real clock.

"Now let's see if you can tell the time on this one!"

"Is it –" Anna looked at the paper clock – "45 minutes past 2?"

"Hooray!" Grandpa was thrilled. "So, if Zane is coming here at 3 o'clock, how long do we have to wait?"

There was a pause. "I KNOW!" said Anna. "15 minutes!"

RR RRRRING

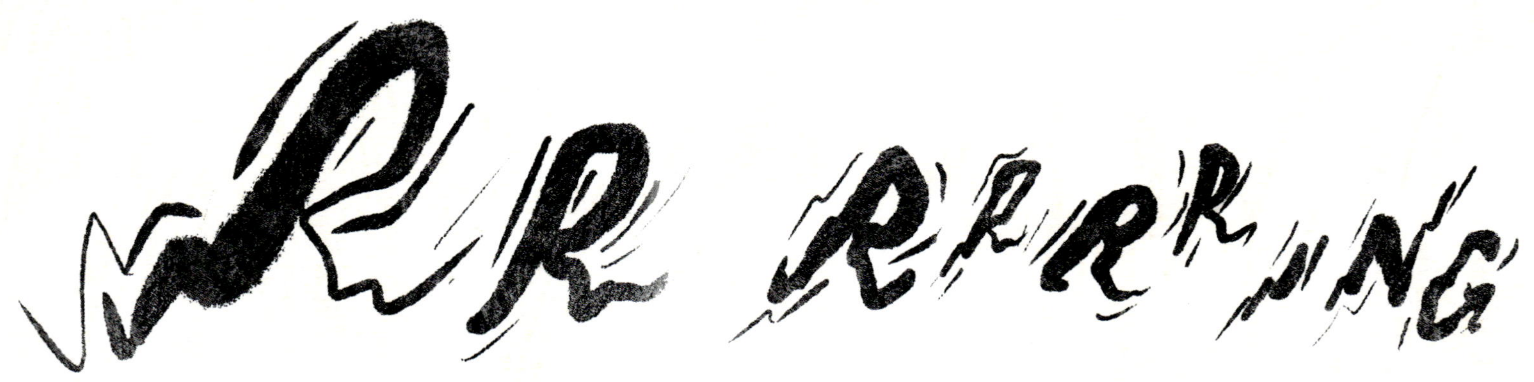

The doorbell made Grandpa and Anna jump.

Anna ran to the door …

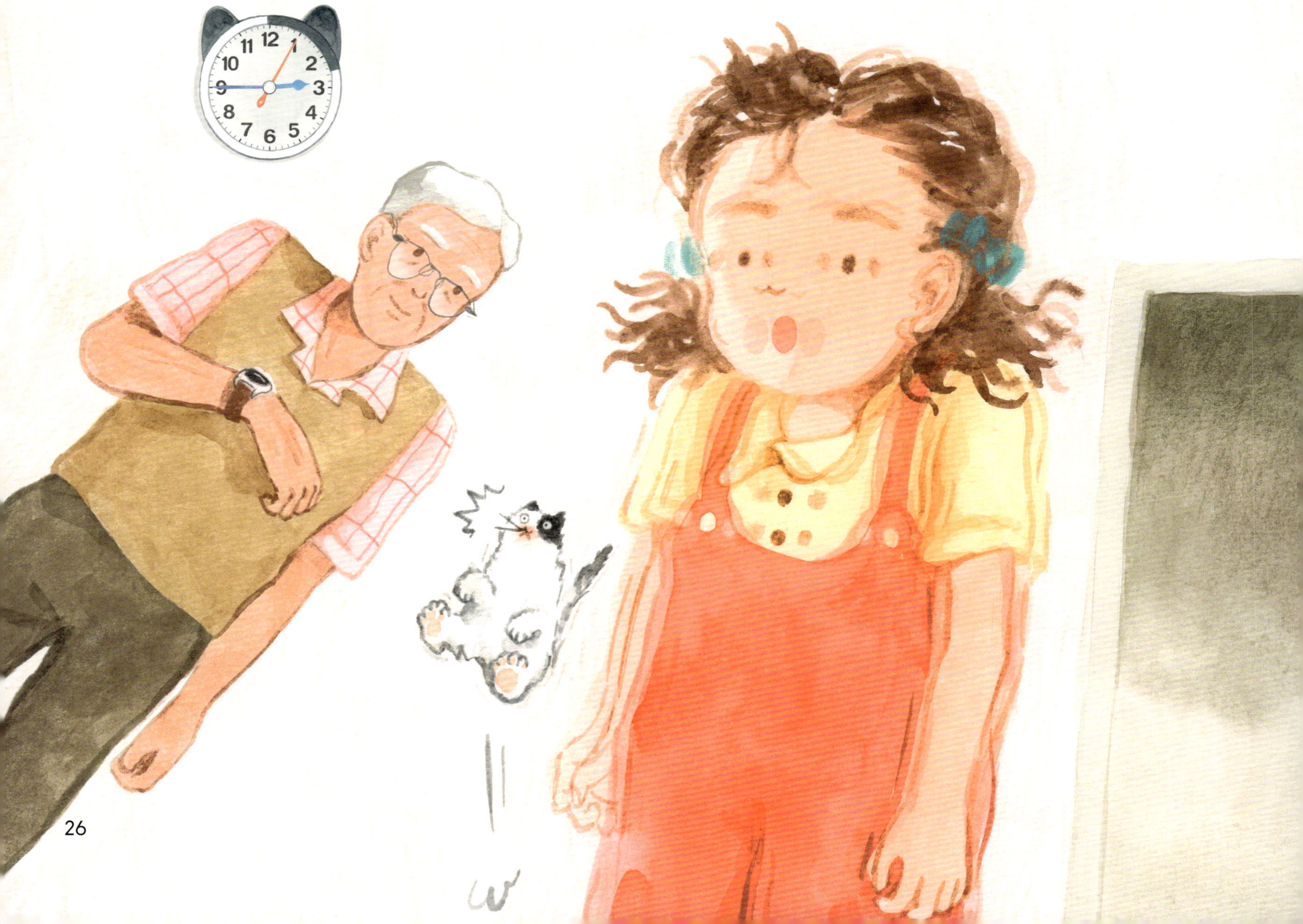

and there was Zane!

"Hello, Zane," she said. "You're 15 minutes early …

but you can come right in!"

A NOTE FROM THE AUTHOR

We can go on measuring time past seconds and minutes and hours. There are **24 hours** in a **day**, and **7 days** in a **week**, and **52 weeks** in a **year**. (If you *really* want to show off, you can tell your friends that there are **1,440 minutes** in a day!)

Do you know *why* a day is 24 hours long? The answer is because it's 24 hours (or 1,440 minutes!) from sunrise to sunrise. If you watch the shadows on a sunny day, you can see them moving … very, very slowly. That's because our planet, Earth, is turning … very, very slowly.

There are lots of other ways you can play and experiment with time, too – here are a few!

1. Ask if you can set a kitchen timer for 1 minute. How many times can you jump up and down? What else can you do in a minute?

2. Now try 5 minutes. Can you get undressed and ready for bed in 5 minutes? What else can you do?

3. If you say "Mississippi" 10 times, that's about 10 seconds. If you say it 60 times, it's a minute. Check it out with the timer! Then try saying "Mississippi" 10 times with a friend. Did you finish at the same time?

4. Why not have a go at making your own time chart? Draw 24 squares on a piece of paper: one for each hour in the day. Then fill in what you'll do in each hour!

INDEX

A.m. and p.m. 22
Clocks 4, 12, 17, 18, 19, 21, 22, 24, 25
Hands 4, 12–13, 14, 15, 18, 19, 21
Hours 11, 16, 18, 21, 22, 28, 29
Measuring 8–9, 12, 28
Minutes 6, 11, 14–15, 16, 18, 19, 20, 21, 25, 27, 28, 29
Seconds 11, 12–13, 15, 16, 28, 29
Time 4, 6, 9, 12, 19, 21, 22, 23, 24, 28, 29

Also in this series:

978-1-5295-1279-3

Available from all good booksellers